Micael Justino

# Proposal for the execution of internal gypsum block masonry

Micael Justino

# Proposal for the execution of internal gypsum block masonry

## Proposal for a standard

ScienciaScripts

Cover image: www.ingimage.com

This book is a translation from the original published under ISBN 978-613-9-68443-4.

Publisher:
Sciencia Scripts
is a trademark of
Dodo Books Indian Ocean Ltd. and OmniScriptum S.R.L publishing group

120 High Road, East Finchley, London, N2 9ED, United Kingdom
Str. Armeneasca 28/1, office 1, Chisinau MD-2012, Republic of Moldova, Europe
Printed at: see last page
**ISBN: 978-620-8-21811-9**

# Summary

*To my parents, Micael and Geliane, for their strength To my sister for her support and to my friend Anna Elis for her example of professionalism.*

# Thanks

To God for giving me the health and strength to overcome difficulties. To this university, its teaching staff, management and administration, who have given me the opportunity to see a higher horizon, made possible by the merit and ethics present here.

To my advisor, Yèda Póvoas, for her support in the short time she had, for her correction and encouragement.

To my parents, for their love, encouragement and unconditional support.

To my friend Anna Elis, who has always been an example of a professional to follow, with all her discipline and dedication to making her dreams come true.

To all those who make up TECOMAT, a company that has contributed in an essential way to the composition of my professional profile, introducing me to and inserting me into the world of the subject addressed below.

# Summary

In the context of sustainable development, where we are increasingly looking for a way to produce on a large scale without losing quality or agility, new technologies are emerging in the construction industry to make this possible. Vertical internal gypsum block walls come with this proposal, since in addition to speed of execution, they offer advantages not found in conventional walls, such as: hydrothermal control of the ambient temperature, acoustic insulation, fine finish, fire resistance, almost zero shrinkage, lightness for structures, easy reconstruction, among others. However, as the technology is new to the market, its execution methods are not yet standardized and not always known. The aim of this work is to develop an execution method that prioritizes the speed and quality of the walls, thus achieving greater productivity (faster x lower cost). The methodology adopted was to compare three types of execution processes, from three different construction companies, with the existing draft standard 02:103.40-014 and thus create a single text. To prepare this text, visits were made to construction sites where fitters were asked about the most important items in the procedure. By standardizing these methods, the aim is to achieve greater control over the execution of walls, so that in the future there will be certainty in the choice of this new technology, so that it can be used on a large scale to replace ceramic block masonry.

**Key words:** New technology. Internal masonry. Gypsum block.

# Chapter 1

## Introduction

### 1.1 Justification

In recent years, the construction industry has made more room for the incorporation of new technologies that can combine speed, economy and sustainability, with the aim of reducing construction costs and maintaining quality standards.

Neves (2011) points out that the masonry of a building has a significant percentage in the total cost of the work, which can reach 20 to 40%, considering the interrelationships with the set of window frames, electrical and hydro-sanitary installations and coverings. According to Marcos Regada Filho[1] in an interview, "weight is synonymous with cost, and since plaster block partition walls are lighter than conventional partitions, it was concluded that the impacts, both on the structure and the foundation, are quite positive in terms of reducing the loads and inputs of materials in the building" (BRASIL GYPSUM MAGAZINE, 2008) and the use of gypsum blocks in internal vertical sealing is an alternative for this cost reduction.

With a view to the prosperity of this technology, a number of companies have begun to specialize and offer the service of erecting Internal Vertical Gypsum Block Fences (VVI-BG).

Each construction company develops its own working method, called the Service Execution Procedure (PES), based on a draft standard 02:013-40-014 (ABNT, 2005) - Execution of plaster block masonry - Procedure.

The procedure developed by the construction companies dictates the erection of the walls, from the conditions for starting the job, to the cladding and, in some cases, even the application of the plaster lining. Once the PES has been developed, the construction company can usually do two things: train its own team to assemble the walls or hire an outsourced contractor to do the job. From there, problems arise, as each company has a different procedure, which doesn't always prioritize practicality when it comes to execution and often has unfeasible requirements that can't be applied in practice. This heterogeneity in the method of execution has consequences such as a reduction in the speed of production, dissatisfaction on the part of professionals who generally earn according to what they produce and difficulties when it comes to changing the workforce, since it will have to be adapted to the reality of the company in question.

It is necessary to carry out a study that makes it possible to standardize these methods of execution, with a view to practicality, safety and feasibility of application, in order to standardize assembly and the profile of the worker.

[1] Partner and commercial director of construction company Franere - Sâo Luiz/MA

## 1.2 Objectives

### 1.1.1 General objective

0 The aim of this work is to propose a procedure for executing internal masonry in plaster block.

### 1.1.2 Specific objectives

To achieve the general objective, the following specific objectives were set:

- analyze different service execution procedures (PES) for internal plaster block masonry to form a single text;

- visiting construction sites to feed back on the text produced, trying to get as close as possible to the reality of the assemblers who make up the workforce in this sector.

The aim of this action is to standardize these methods with a view to greater practicality, safety and feasibility of application, so that assembly and the technical profile of the worker can be standardized.

### 1.3 Structuring the work

Chapter 2 describes the literature review on the characterization of the materials and components of gypsum block masonry, as well as the advantages of its use.

Chapter 3 contains the methodology developed in the work.

Chapter 4 contains the proposed standard for the execution of internal gypsum block masonry.

Chapter 5 contains the conclusions followed by the references used in the work.

The original text of the proposed standard can be found in the appendix.

# Chapter 2

## Gypsum block masonry

### 2.1 General about blocks

According to SINAT (2012) and SENAI (2013), gypsum block masonry used for vertical sealing can form the internal walls of single-family single-storey houses, single-family townhouses, including overlapping houses or multi-storey multi-family housing buildings, commercial and industrial buildings, schools and hospitals. These walls have no structural function or function as the building's external wall.

A positive factor for the use of these blocks is that the country's largest gypsum reserve (the raw material for gypsum) is located in the Araripina Gypsum Complex, facilitating the sale of gypsum blocks in the North and Northeast regions.

According to data provided by SUPERGESSO[1] , the Araripe Gypsum Complex has 1.22 billion tons of gypsum and is expected to last 600 years. 30% of the country's gypsum reserves are located in the complex, which currently produces 98% of the gypsum consumed in Brazil. It can also be said to be one of the largest and most important reserves in the world, due to the high purity of the gypsum supplied by the region.

With one of the largest gypsum reserves located in Pernambuco, this state has another advantage: lower freight costs.

The Araripina Plastering Pole is located in the epicenter of the semi-arid Northeast, about 800 km from seven Brazilian capitals (Recife, Salvador, Fortaleza, Aracajù,

1 Gypsum block factory, located in Araripina, with office in Boa Viagem, Recife- PE

Maceiò, Joâo Pessoa and Natal). Occupying 8% of Pernambuco's territory, it covers the municipalities of Araripina, Bodocó, Ipubi, Ouricuri and Trindade and has around 230,000 inhabitants.

## 2.2 Characterization of materials and components

### 2.1.1 Blocks

According to SINAT (2012) and SENAI (2013), the characteristics of gypsum blocks can be determined according to: color, density, dimensions, strength, PH and the presence or absence of voids. There are four types of gypsum blocks on the market:

Standard gypsum block - S: The standard gypsum block, presented in white, should preferably be used in the construction of internal partition walls in dry areas. Made from gypsum and water, the characteristics of these blocks are shown in Table 1.

Table ICharacteristics of Standard-S gypsum blocks

| Features | Bioco thickness in mm | | | |
|---|---|---|---|---|
| | 70 | 70 | 100 | Tolerance |
| Type | Leaked | Massive | Massive | - |
| Dimensions in mm | 666x500 | 666x500 | 666x500 | 0,5 |
| Average weight in Kg of a bioco | 19 | 24 | 34 | 5% |
| Average weight in $KgZm^2$ | 54 | 72 | 102 | 5% |
| Surface hardness in Shore C | >=55 | >=55 | >=55 | - |
| Flexural strength modulus (MPa) | >=2.0 | | | - |
| Compressive Strength (MPa) | >=3,0 | | | - |

Source: SINAT (2012)

Water-repellent gypsum block - HIDRO: Water-repellent gypsum block, presented in blue, should preferably be used in the construction of internal partition walls in wet and dry areas. It is usually placed in the first row of walls in dry areas that are periodically washed. It consists of special plaster, water and a water-repellent additive.

The characteristics of these blocks are shown in Table 2.

Table 2.Characteristics of Hydrogenated gypsum blocks-HIDRO

| Features | Block thickness in mm | | | |
|---|---|---|---|---|
| | 70 | 70 | 100 | Tolerance |
| Type | Leaked | Macico | Massive | - |
| Dimensions in mm | 666x500 | 666x500 | 666x500 | 0,5 |
| Average weight in Kg of a bioco | 19 | 24 | 34 | 5% |
| Average weight in Kg/m$^2$ | 54 | 72 | 102 | 5% |
| Surface hardness in Shore C | >=55 | >=55 | >=55 | - |
| Water absorption | <5% | <5% | <5% | - |
| Flexural strength modulus (MPa) | >=2,0 | | | - |
| Compressive Strength (MPa) | >=3,0 | | | - |

Source: SINAT (2012)

Glass Fiber Reinforced Gypsum Block - GRG: presented in green, it should preferably be used in the construction of internal wall areas, dry areas, which require greater resistance to pulling, greater bending and in safety areas for escape in the event of fire. The characteristics of these blocks are shown in Table 3.

Table 3.Characteristics of glass fiber reinforced gypsum blocks-GRG

| Features | Block thickness in mm | | | |
|---|---|---|---|---|
| | 70 | 70 | 100 | Tolerance |
| Type | Leaked | Massive | Massive | - |
| Dimensions in mm | 666x500 | 666x500 | 666x500 | 0,5 |
| Average weight in Kg of a block | 19 | 24 | 34 | 5% |
| Average weight in Kg/m$^2$ | 54 | 72 | 102 | 5% |
| Surface hardness in Shore C | >=55 | >=55 | >=55 | - |
| Flexural strength modulus (MPa) | >=3,0 | | | - |
| Compressive Strength (MPa) | >=3,0 | | | - |

Source: SINAT (2012)

GRGH - Glass Fiber Reinforced and Hydrated Gypsum Block: presented in pink, should preferably be used in the construction of external or internal wall areas, wet and damp areas, which require greater resistance to pulling and bending. The characteristics of these blocks are shown in Table 4.

Table 4.Characteristics of Glass Fiber Reinforced Gypsum Blocks with Hydrofugation-GRGH

| Features | Bioco thickness in mm |
|---|---|

| | 70 | 70 | 100 | Tolerance |
|---|---|---|---|---|
| Type | Leaked | Macico | Massive | - |
| Dimensions in miri | 666x500 | 666x500 | 666x500 | 0,5 |
| Average weight in Kg of a bioco | 19 | 24 | 34 | 5% |
| Average weight in Kg/m$^2$ | 54 | 72 | 102 | 5% |
| Surface hardness in Shore C | >=55 | >=55 | >=55 | - |
| Water absorption | <5% | <5% | <5% | - |
| Flexural modulus of resistance (MPa) | >=3,□ | | | - |
| Compressive Strength (MPa) | >=3,0 | | | - |

Source: SINAT (2012)

### 2.1.2 Plaster glue

According to the SINAT Guidelines (2012), gypsum adhesive is made from gypsum and additives. When mixed in the appropriate water/gypsum glue ratio, 20 kg (01 bag) of gypsum glue to 13 liters of water, it has a pasty consistency that allows it to be applied with tubes, spatulas or similar tools. It is a powdered product, supplied in bags of 20, 5 and 1 kg, designed to be used in the assembly of horizontal (walls) and vertical (ceilings and ceilings) sealing systems built with precast gypsum.

It can also be used for bonding other plaster elements such as: crown moldings, moldings, boards, plasterboard panels, for bonding tiles, ceramics and tiles. The characteristics of this adhesive plaster are shown below.

Gesso glue Standard - S: white in color, it should preferably be used to glue gypsum blocks used in internal partition walls in dry areas. However, for many years it was used as the only product to fix the blocks.

Water-repellent adhesive plaster - HIDRO: blue in color, it should preferably be used for gluing plaster blocks used in internal partition walls in wet and damp areas.

The only difference between the two adhesives is their water absorption. While Standard adhesive plaster has an absorption of between 30% and 40%, Hydro adhesive plaster has values of less than 5%. Data was obtained from the SINAT Guideline (2012).

## 2.3 Advantages of using gypsum blocks

As it is a new technology and not so widespread in the domestic market, the use of gypsum blocks to replace traditional ceramic blocks in the production of internal walls has been increasingly studied in order to find advantages that justify this replacement. According to Lordsleem Jr (2011), some of these advantages can already be highlighted: shorter construction time for seals, greater floor area, the possibility of installation over the final floor, greater layout flexibility, lighter walls, less overload on structures and foundations, greater dimensional accuracy and better thermo-acoustic comfort.

### **2.3.1** Shorter processing time

When comparing gypsum blocks with conventional masonry, the following observation can be made: a gypsum block has an average size of 0.33m$^2$ which is equivalent to 6 bricks (8-hole ceramic block). As a result, 1m2 of conventional masonry walling requires around 20 blocks, while gypsum walling requires only 3.

Another factor that interferes with production viability is the number of services required to execute the systems. Table 5 shows a comparison between the two systems.

Table 5.Comparison between internal masonry of ceramic block and gypsum block

| Systems | Services involved | Productivity (hour/m )$^2$ | Finished service time (hour/m2) |
|---|---|---|---|
| Ceramic block masonry | > Elevation | > 0,45 | |
| | > Chapisco | > 0,08 | |
| | > Plaster | > 0,45 | 1,13 |
| | > Trimming and painting | > 0,15 | |
| Gypsum block masonry | > Elevation | > 0,36 | |
| | > Pipe laying | > 0,20 | 0,68 |
| | > Trimming and painting | > 0,12 | |

Source: Pires Sobrinho et al. (2009)

### 2.3.2 Stronger walls

According to Pires Sobrinho et al. (2009), gypsum block masonry is lighter, ranging from 0.6 $kN/m^2$ to 1.0 $kN/m^2$ , when compared to traditional ceramic block masonry coated with mortar, which ranges from 1.2 $kN/m^2$ to 1.8 $kN/m^2$ , contributing to a reduction in permanent loads on the slabs/beams. This reduction also influences the reduction of immediate deflections and slow deformation of concrete slabs.

### 2.3.3 Thermo-acoustic comfort

Plaster elements or coverings can help to improve the soundproofing of rooms in various ways. Peres, Benachour and Santos (2008) cite the following examples:

Thanks to the continuity of the coatings over traditional masonry, they are able to fill in all the possible cracks and holes through which sound propagates;

Thanks to its plasticity, plaster can be used to make decorative elements with a specific geometry capable of suppressing or attenuating the reverberation of sounds emitted into the room where the plaster elements are installed.

### 2.3.4 Fire resistance

Also according to Peres, Benachour and Santos (2008), coatings and elements made of gypsum, alone or in combination with other materials, significantly improve the thermal insulation of walls due to their low coefficient of thermal conductivity. This coefficient, which in the specific case of gypsum varies with humidity and the density of the hydrated and dry material, is in the order of 0.25 to 0.50 according to the conditions shown in Table 6.

Table 6 - Variation of the thermal conductivity coefficient according to the density of gypsum-based materials

| Characteristics of gypsum-based materials | Density (Kg/m )$^3$ | Conductivity (W/m$^0$ c) |
|---|---|---|
| Plasters incorporating air or other elements | 600 a 900 | 0,30 a 0,25 |
| Pre-molded gypsum elements such as boards, blocks and others. | 900a1000 | 0,35 |
| Gypsum mortars for | 1100 a 1300 | 0,50 |

| | | |
|---|---|---|
| projection and high hardness plasters | | |

Source : Peres, Benachour and Santos (2008)

# Chapter 3

## Methodology

For this research, a bibliographical study was carried out on the subject of plaster block masonry in order to obtain greater knowledge and update on the technological advances of this service.

Three (3) types of PES were obtained from different companies. These documents served as the basis for the text of the draft standard

02:013-40-014 (ABNT, 2000) - Execution of plaster block masonry - Procedure.

The text was discussed by a support group created specifically for this procedure, with a view to making adjustments to the standard to be submitted for national consultation.

After the group discussion, the production of the walls on site was monitored to observe the feasibility of the procedures in practice.

Three construction sites were visited: two of them in Ipojua, on the seafront beaches of Cupe and Muro Alto, and the last one in the Boa Viagem neighborhood of Recife.

Subsequently, a feedback process was carried out on the text produced in order to present a proposal for the execution procedure of plaster block masonry.

# Chapter 4

## Results

### I Project 02:103.40-014

### Execution of sealing masonry in gypsum blocks

Procedure

---

Keyword(S): Plaster, plaster bioco, masonry.

Summary

Preface

Objective

Normative references

Definitions

Preliminary conditions

General conditions

Specific conditions

Inspection

Preface

ABNT - the Brazilian Association of Technical Standards - is the National Standardization Forum. The Brazilian Standards, whose content is the responsibility of the Brazilian Committees (ABNT/CB) and the Sectoral Standardization Bodies (ONS), are drawn up by Study Commissions (ABNT/CE), made up of representatives of the sectors involved, including: producers, consumers and neutrals (universities, laboratories and others).

The draft Brazilian standards drawn up by the ABNT/CB and ONS are circulated for public consultation among ABNT members and other interested parties.

Objective

This standard determines the procedures required for the execution and inspection of plaster block masonry.

## 3 Normative reference

The standard listed below contains provisions which, when cited in this text, constitute prescriptions for this Standard. The edition indicated was in force at the time of publication. As every standard is subject to revision, it is recommended that those entering into agreements based on this standard check the advisability of using the most recent edition of the standard cited below. ABNT has information on the standards in force at any given time.

NBR 1313: - Modulated internal lightweight partitions - Specification

NBR 6494: - Scaffolding safety procedure

Basic Text - 02:002.40.10 - Gypsum blocks for masonry - Specification

Base Text - 02:002.40.13 - Plaster glue - Specification

## 4 Definitions

**1.1** Girder: structural component located over the masonry beams.

**1.2** Vain: existing opening in masonry

**1.3** Counter-span: structural component located under the masonry spans.

**1.4** Frames: wooden components or metal and/or plastic profiles that have the function of structuring the frame elements.

**1.5** Alizar: a component which, in conjunction with the frames, covers the interface between the frames and the masonry.

**1.6** Movement joint: system of rigid discontinuity and/or flexible continuity between masonry elements and masonry elements and structure or window frames.

**1.7** Skirting board: a component that, together with the structure or masonry, covers up movement joints.

**1.8** Bay: a wooden component whose function is to serve as a support element for fixing skirting boards and/or aligning the structure and/or window frames.

**1.9** Slightly stiff structural element: structural elements that have the possibility of deformation, under expected loading, greater than L/500 in span and H/250 in height, considering L (width) and H (height) in cm.

**1.10** Tie joint: a system for laying all masonry components in which the vertical joints are discontinuous.

**1.11** Connection: the link between masonry and structural components (pillars, beams, etc.) achieved through the use of particular materials and construction arrangements.

**1.12** Laying joint: the meeting of the fitting joints for assembling the masonry components.

**1.13** Gypsum glue: a specific powder product to be used for gluing gypsum components.

# 5 Preliminary conditions

## 5.1 Conditions for starting work

**5.1.1** . The pillars and beams that will meet the gypsum block masonry must be plastered at least 72 hours before the job.

**5.1.2** . The main axes must be defined according to the project and located or transferred to the working floor, as well as the frames and door grilles.

**5.1.3** . The power supply cables for the equipment used on the floor must be suspended from an electrical panel fitted with an IDR protection device (in accordance with NR10).

**5.1.4** The subfloor must be clean/cleaned at the location of the 1st layer[a] and the plaster brushed at the locations where the fences meet.

**5.1.5** The work area must be clear.

**5.1.6** If there is peripheral and internal masonry using another material, this must be completed.

**5.1.7** Scaffolding must meet the requirements of NBR 6494.

## 5.2 Leveling the floor and making the structure plumb and square

The leveling of the floor and the plumbness and squareness of the structure must be checked in the areas specified in the architectural project for the execution of masonry. The tolerances for the structure must comply with NBR 14931. A maximum deviation of 0.5% is allowed in the leveling of the floor.

# 6 General conditions

## 6.1 Masonry execution

**6.1.1** Masonry components must meet the specifications contained in draft standard 02:003.02.10.

**6.1.2** Before the block is laid, it must be cleaned with a nylon brush on the sides that will receive the adhesive plaster, to ensure perfect bonding.

**6.1.3** The position of the blocks and the details for tying the joints must comply with the executive project and/or the architectural, installation and structural project.

**6.1.4** Start the marking following the Executive Project for the fences, stretching the lines of the main axes, indicating the type of block used in each wall and the possible openings.

**6.1.5** Prepare the adhesive plaster following the manufacturer's recommendations.

6.1.5.1 For the water-repellent block, use water-repellent adhesive plaster.

6.1.5.2 Plaster adhesive can be used at the interface between the water-repellent block and the standard block.

6.1.5.3 At the interface of the water-repellent block with conventional seals and concrete elements, use water-repellent adhesive plaster, wetting the areas of conventional seals and concrete elements beforehand.

**6.1.6** Run 1 row.[a]

6.1.6.1 The blocks in the first row of ties must be waterproofed with the female edge downwards, previously filled with adhesive plaster.

6.1.6.2 The 1st row should begin with a plaster block being placed against a masonry or pillar that will be glued to the subfloor with plaster adhesive, and on the vertical side to the wall or pillar. Continue assembling the first row, always applying plaster adhesive to the part that will be glued to the floor and to the plaster block that has already been fixed.

6.1.6.3 Apply o gypsum glue to the vertical and horizontal joints, at the interface between the block and the pillar/periphery masonry, letting the glue run down the joints between the blocks and between the blocks and the floor, removing the excess afterwards. Press the block firmly against the surface of the floor and the other joints using a rubber hammer.

6.1.6.4 As the blocks are laid with the help of a rubber hammer, the adhesive plaster must flow through the laying joints. The joints must be at least 2mm thick and continuous.

**6.1.7 Run the 2nd[a] row.**

6.1.7.1 A2[a] row should always be started on the side where the first row began, using a cut block, ensuring that the vertical joints are at least 20 cm apart.

6.1.7.2 The masonry must be modulated in such a way as to use the greatest number of whole components, using discontinuous vertical joints and positioned in such a way as to optimally meet the installation and architectural projects.

6.1.7.3 Hard joints between gypsum block masonry components and between masonry elements and the structure must be made continuously. Plaster adhesive must be used.

6.1.7.4 The modulations of the components in the connection joints in the masonry elevation must follow the indications shown in Figure 1.

| **1** Walls that cross each other with transom ties. | **2** Walls that meet with alternating ties. | **3** Walls that meet other vertical elements with end ties. |
|---|---|---|
| 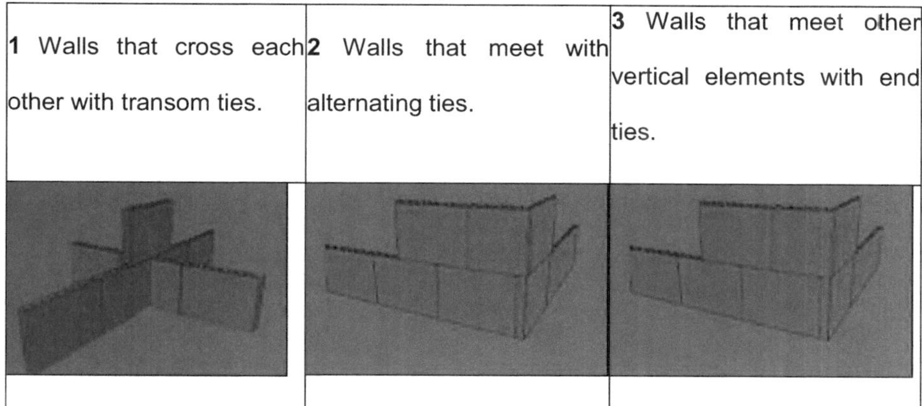 | | |
| | | |

Figure 1 - Component modulations at the connection joints in the masonry elevation

### 5.2 Execution of meetings

When buildings are built, gypsum block masonry is also connected to other walls or building elements, such as slabs, beams and pillars. These joints require a technique for their execution.

5.2.1 When joining masonry components to framing elements, additional components must be used to ensure fixing at discrete points, as well as using materials that continuously fill the gaps in the connection.

5.2.2 When materials with different expansion moduli meet vertically (gypsum masonry x concrete structure and/or ceramic block masonry) it is necessary to use mesh during execution. The mesh must be applied at least 5 cm on each side and along the entire length of the intersection of the materials.

5.2.3 When laying the last row of blocks, leave a gap of 2.0cm to 3.0cm between the top of the block and the meeting element, which will be filled in at least 24 hours after the masonry has been assembled.

5.2.4 When the plaster blocks meet the beams, the surface of the beams should be grouted and a strip of EPS (expanded polystyrene) glued to it. The width of the EPS must exceed the thickness of the block by 1cm on each side (the thickness of the cladding) and be glued to the slab with plaster adhesive. The EPS must have a density of between 20 and 23 kg/m$^3$ and be at least 10 mm thick. The last block should be chamfered to make it easier to place the filler.

5.2.5 The remaining space between the top of the seal and the EPS strip should be filled with a mixture of adhesive plaster + casting plaster (1:1), placing the paste on the open side.

5.2.6 Masonry tightening should be delayed as much as possible. The ideal condition is for the elevation phase to be completely finished or for at least 3 upper floors to be completed.

### 5.3 Door seals

5.3.1 The gaps in the door grilles must be left open when the masonry is lifted and the meeting of the blocks in the middle of the gap must be reinforced with polyester mesh.

5.3.2 Precast lintels should be placed over openings, doors and windows and when the opening is greater than 2.40m the concrete lintel should be calculated as a beam. If the lintel is made of plasterboard, it must have a minimum height of 15 cm.

5.3.3 When precast lintels are used, they must be centered on the sealing axis and manufactured 2.0 cm smaller than the thickness of the block, with a thickness of 1 cm on each side waiting for the cladding. Their height must be a maximum of 10.0 cm and exceed the width of the gap by at least 20.0 cm on each side.

5.3.4 The polyester mesh should be applied 5 cm to each side and along the entire length of the intersection of the materials.

5.3.5 Door and door grilles can be fixed with expanded polyurethane. For the best performance of this form of fixing, some precautions are necessary.

5.3.5.1 Bring the plaster block seal as close as possible to the door frame so that the door fits tightly, in order to comply with NBR 15930.

5.3.5.2 Thoroughly clean the block cavities and grids at the meeting point and spray the surfaces where you want the expanded polyurethane to adhere with clean water.

5.3.5.3 Apply the polyurethane to the entire perimeter of the railing and spray clean water over the applied polyurethane. Wait for the foam to fully cure before finishing the sides by cutting off the excess polyurethane.

5.3.5.4 The smoothing must perfectly cover the wall/door grid connection.

5.4 **Facilities**

5.4.1 Pipes and electrical boxes must be installed 24 hours after the masonry has been fixed. The tears in the walls must be made with an electric machine so as not to damage the stability of the blocks.

5.4.2 When hollow blocks are used, the conduits must be inserted into the holes in the blocks and, therefore, these seals must have a specific design that can include the placement of blocks in the horizontal position (the larger side parallel to the floor slab), in the vertical position (smaller side parallel to the floor) or mixed (contemplating both positions) in order to reduce as much as possible the need to cut the seals to pass conduits after the wall has been completed.

5.4.3 When compact blocks are used, after the joints and connections in the walls have completely dried, the cuts can be made with special notching machines, allowing the depth of the cut to be precisely regulated. The pipes must be covered by a minimum of 4mm and must be covered with a mixture of foundry plaster and adhesive plaster (1:1).

5.4.3.1 The maximum diameter of the conduit to be inserted must be less than 1/3 of the thickness of the block and with a coating greater than or equal to 4mm thick. The width of the cut must be 2mm more than the outer diameter of the conduit to be inserted (Figure 2).

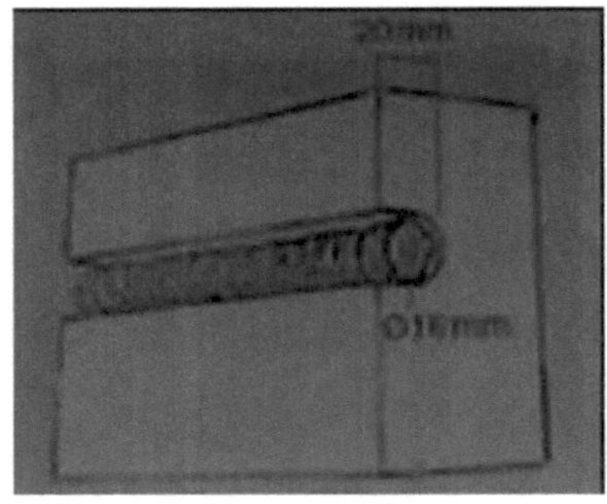

Figura 2

5.4.3.2 The pit should be brushed vigorously and all loose material removed. Insert the conduit into the pit up to the bottom, fill 2/3 of the pit volume with a mixture of adhesive plaster and foundry plaster (1:1) (Figure 3).

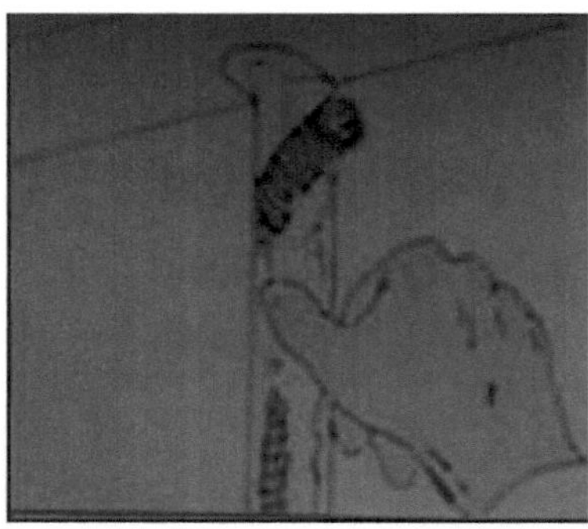

Figura 3

5.4.3.3 Close the pit and finish the surface with a mixture of plaster for casting and plaster adhesive (1:1) (Figure 4).

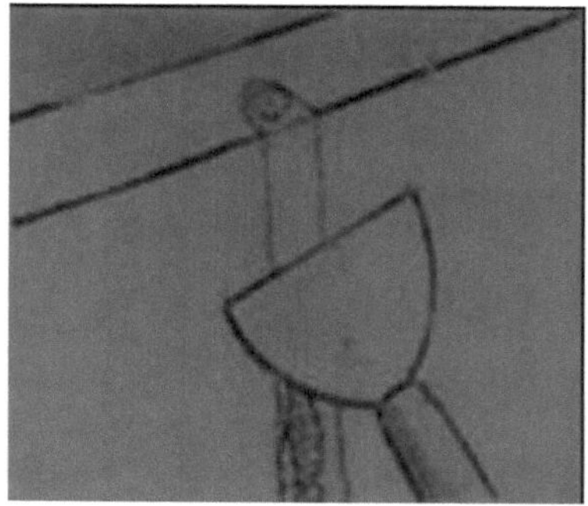

Figura 4

5.4.4 Where connectivity/electricity blocks are installed, we recommend using 100 mm thick gypsum blocks. As the possibility of splitting the gypsum block masonry is high, it is recommended that the finishing in this area be done with gypsum adhesive mortar.

5.4.4.1 To make the mortar reinforced with adhesive plaster, place the polyethylene mesh in the area of the cut and recompose the section with the adhesive plaster. Do not use hydro-glue plaster.

5.4.4.2 After the section has been recomposed, the polyester mesh is laid over the distribution box and the conduits with a 5 cm gap on all sides.

**7 Specific conditions**

**6.1 Wet areas**

**6.1.1** In wet areas, where there is water on the walls, it is recommended to use hydrophilic components throughout the area.

**6.1.2** In areas subject to capillary water rise, the use of waterproofing devices at the base is recommended. Figure 5 shows some of the possible types of insulation.

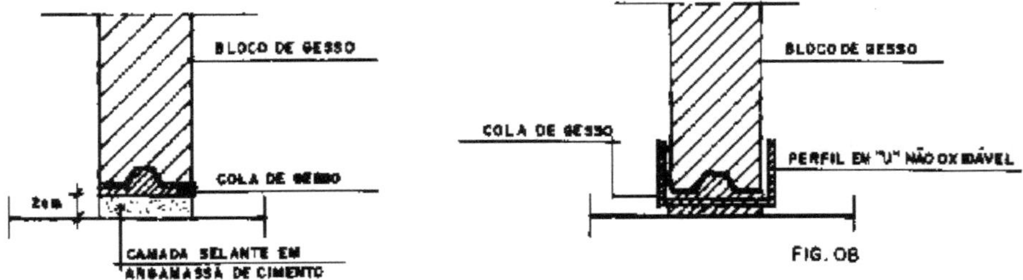


Figure 5 - Possible types of insulation

## 6.2 Weak structures

1.1.1 In masonry built on weak structural elements, it is recommended that devices be used to absorb deformations, such as the use of elastic joints, within the specified limits, keeping the masonry stable and free of cracks. The joints can be made of cork, felt, expanded polystyrene, expanded rubber or other materials suitable for this purpose and must be made with adhesive plaster. Figures 6, 7 and 8 show some possible devices for use at the joints with vertical elements (walls and/or pillars) and horizontal elements (floor, slab and beam).

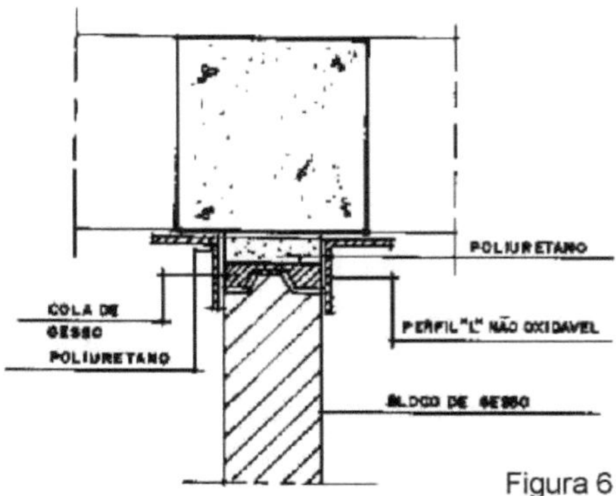


Figura 6

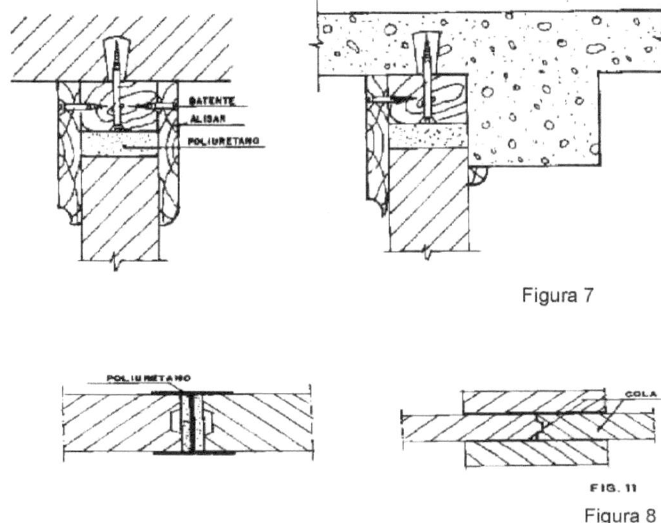


Figura 7

Figura 8

1.1.2 When building masonry in long spans (L > 5m), it is recommended to use structuring devices that guarantee performance in accordance with the specifications of draft standard 02.002.40.12.

1.1.3 When executing masonry with a height greater than 3m, it is recommended to use structuring devices that guarantee performance in accordance with the specifications of draft standard 02.002.40.12.

**6.3 Where the plaster cladding of the enclosing masonry walls and/or structures has been completed**

6.3.1 The cladding layer should be removed so that the gypsum block masonry can be fixed directly to the ceramic block/frame, ensuring better fixation.

6.3.2 In joints and complementary plaster walls, at least 20cm of the applied block must be removed, alternately, to ensure that the blocks are correctly tied together.

## 6.4 Fixing suspended parts

The fixing of suspended parts must comply with NBR 15575.

## 8 Inspection

**7.1** It is the responsibility of the site supervisor to inspect and take delivery of the masonry. All masonry must be inspected in accordance with the criteria set out in this standard.

## 7.2 Rental

7.2.1 It must be checked before the masonry is assembled and verified after the masonry has been assembled, and it must be in accordance with the dimensions of the specific project.

7.2.2 Instruments with the precision of tape measures and building squares can be used for this check.

## 7.3 Wall flatness

7.3.1 It must be checked during the assembly of the masonry and verified after the masonry has been assembled, with no distortion greater than 5 mm.

7.3.2 A metal ruler can be used to check this by positioning it at various points on the wall.

## 7.4 Plumb

It must be checked during the assembly of the masonry and verified after the masonry has been assembled.

## 7.5 Level

It must be checked during the assembly of the masonry and verified after the masonry has been assembled. This check can be made using a transparent plastic hose with a diameter greater than or equal to 13 mm.

# Chapter 5

## Final considerations

The lack of standardization of gypsum-based products, as well as the low quality of some manufacturers due to their almost entirely handmade production and, above all, the lack of skilled labor, are some of the problems faced by the national gypsum industry. These problems stifle the potential consumption of gypsum-based products by the construction market in Brazil.

Creating a single text that can complement a draft standard, which at the moment is seen as insufficient to meet the needs of the market, can bring us confidence in such technology that will enable two major advances in this area: bringing confidence in the subject for its application on a large scale and changing the national panorama that says there is no specialized workforce in Brazil.

# References

BRAZILIAN ASSOCIATION OF TECHNICAL STANDARDS. **NBR 6118.** Design of concrete structures - procedure. Rio de Janeiro, 2007.

. **Draft standard 02:013-40-014.** Execution of plaster block masonry. Rio de Janeiro, 2000.

NEVES, M.L.R. **Construction method for internal vertical sealing with gypsum blocks.** Master's degree dissertation, Escola Politècnica de Pernambuco, Recife, 2011.

PERES, L.; BENACHOUR, M.; SANTOS, V.A.dos. **Gypsum: production and use in civil construction.** 120p. SEBRAE. Recife, 2008.

PIRES SOBRINHO, C. W. A.; BEZERRA, N. M.; COSTA, T.C.T.; SILVA, C.B.A. **Internal partitions in gypsum block masonry buildings: technical, economic and environmental advantages.** .ITEP. Recife, 2009.

SINAT. National Technical Assessment System. **Guidelines for the Technical Evaluation of Products: Internal vertical walls in non-structural masonry of gypsum blocks**. Brasilia, 2012.

BRAZIL GYPSUM. **Advantages of gypsum blocks**. Available at <http://www.braziliangypsum.com/vantagens.asp>. Accessed on: 02 Apr. 2013.

BRAZIL GYPSUM MAGAZINE. **The voice of construction.** p.7. Editora NE BRASIL. Recife, 2008.

LORDSLEEM JR., A.C. **Execution and inspection of rationalized masonry**. Sâo

Paulo: O Nome da Rosa, 2000. 104 p.

TECNE REVIEW. **Internal vertical sealing of buildings with gypsum blocks**. Available at <http://techne.pini.com.br/engenhariacivil/181/artigo287933-1.aspx>. Accessed on: 02 Apr. 2013.

COSTA, A.M.; INOJOSA, A.C. Gypway construction system. **Gypsum block masonry.** SIDUSGESSO. Construction and maintenance manual. [2013]

CIARLINI, A.G.C.; PINTO, D.C.; OSORIO, A.P. **Plaster: technology that reduces loads and costs in civil construction.** In: ENCONTRO NACIONAL DE ENGENHARIA DE PRODUÇÀO, 2005, Sâo Paulo. **Proceedings...** Salvador, ABEPRO, 2005.

LORDSLEEM JR., A.C. **Execution and inspection of rationalized masonry**. Sâo Paulo: O Nome da Rosa, 2000, 104 p.

ROCHA, C. A. L. **Gypsum in the Construction Industry: Economic Considerations on the Use of Gypsum Blocks.** Dissertaçâo (mestrado), 103 p, Universidade Federal de Pernambuco, Recife, 2007,

PERES, L.; BENACHOUR, M.; SANTOS, V.A.dos. **Gypsum: production and use in civil construction.** 120p. SEBRAE. Recife, 2008.

# Appendices

Nov/2000 **Project 02:103.40-014**

**Execution of plaster block masonry**

Procedure

ABNT/CB-02 - Brazilian Civil Construction Committee

02:103.40 - Study Committee on Natural Gypsum for Civil Construction Project

02:103.40-014 - Performance of slopping brickword in gypsum blocks - Procedure

Descriptors: Gypsum. Gypsum blocks. Slopping.

Keyword(s): Plaster, plaster block, masonry. 5 pages

**Summary**



**Foreword**

ABNT - the Brazilian Association of Technical Standards - is the National

Standardization Forum. The Brazilian Standards, whose content is the responsibility of the Brazilian Committees (ABNT/CB) and the Sectoral Standardization Bodies (ONS), are drawn up by Study Commissions (ABNT/CE), made up of representatives of the sectors involved, including: producers, consumers and neutrals (universities, laboratories and others).

The draft Brazilian standards drawn up by the ABNT/CB and ONS are circulated for public consultation among ABNT members and other interested parties.

## 9 Objective

This standard determines the procedures required for the execution and inspection of plaster block masonry.

## 10 Normative reference

The standard listed below contains provisions which, when cited in this text, constitute prescriptions for this Standard. The edition indicated was in force at the time of publication. As every standard is subject to revision, it is recommended that those who carry out agreements based on this standard check the suitability of using the most recent edition of the standard cited below. ABNT has information on the standards in force at any given time.

NBR 1313: - Modulated internal lightweight partitions - Specification

NBR 6494: - Scaffolding safety procedure

Basic Text - 02:002.40.10 - Gypsum blocks for masonry - Specification

Base Text - 02:002.40.13 - Gypsum glue - Specification

## 11 Definitions

11.1 **Girder:** structural component located over the masonry **beams**.

11.2 **Vain:** existing opening in masonry

11.3 **Counter-span:** structural component located under the masonry spans.

11.4 **Frames:** wooden components or metal and/or plastic profiles that have the function of structuring the frame elements.

11.5 **Alizar: a** component that, together with the frames, covers the interface between the frames and the masonry.

11.6 **Movement joint:** system of rigid discontinuity and/or flexible continuity between masonry elements and/or masonry elements and structure and/or frames.

11.7 **Skirting board: a** component that, together with the structure or masonry, covers up movement joints.

11.8 **Battens:** wooden component that serves as a support element for fixing skirting boards and/or aligning the structure and/or frames.

11.9 **Slightly stiff structural element:** structural elements that can deform more than L/500 in span and H/250 in height under expected loading.

11.10 **Tie joint: a** system for laying all masonry components in which the vertical joints are discontinuous.

11.11 **Connection: the link** between masonry and structural components (pillars, beams, etc.) achieved through the use of particular materials and construction arrangements.

11.12 **Laying joint:** the meeting of the fitting joints for assembling the masonry components.

11.13 **Plaster glue: a** powdered product specifically for use in gluing plaster components.

## 12 Preliminary conditions

### 12.1 Leveling the floor and squaring the structure

The leveling of the floor and the plumbness and squareness of the structure must be checked in the areas provided for in the architectural project for the execution of masonry. A maximum deviation of 0.5% is allowed in each of the checks carried out.

## 13 General conditions

### 13.1 Masonry execution

13.1.1 The masonry components must meet the specifications contained in the basic text 02:003.02.10.

13.1.2 Before laying the block, the faces that will receive the adhesive plaster must be cleaned with a nylon brush to ensure perfect bonding.

13.1.3 The position of the blocks and the details for tying the joints must comply with the executive project and/or the architectural, installation and structural project.

13.1.4 Start the marking following the Executive Project for the fences, stretching the lines of the main axes, indicating the type of block used in each wall and the possible openings.

13.1.5 The modulations of the masonry components in the connection joints in the elevation of the masonry, in the corner bracing and in the bracing of junction and cross walls must follow the indications shown in figures 1,2,3,4 respectively:

13.1.6 Hard joints between masonry components and between masonry elements and the structure must be made continuously. It is recommended to use gypsum glue, or a compound that guarantees stability and performance, as long as it meets the specifications Basic Text 02:002.40.12

Figure 1 - Continuity lashing

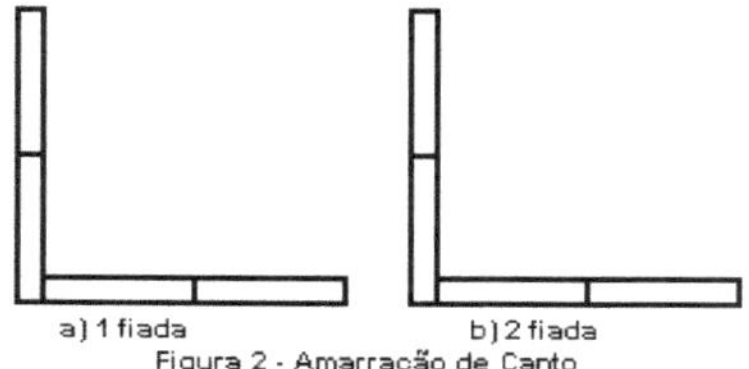


Figure 2 - Corner brace

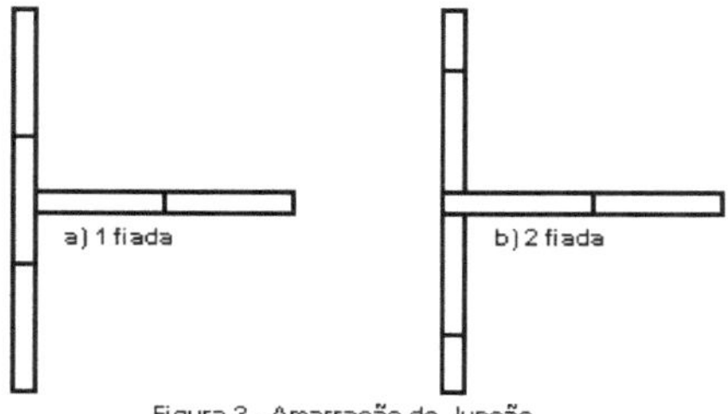


Figure 3 - Junction tie

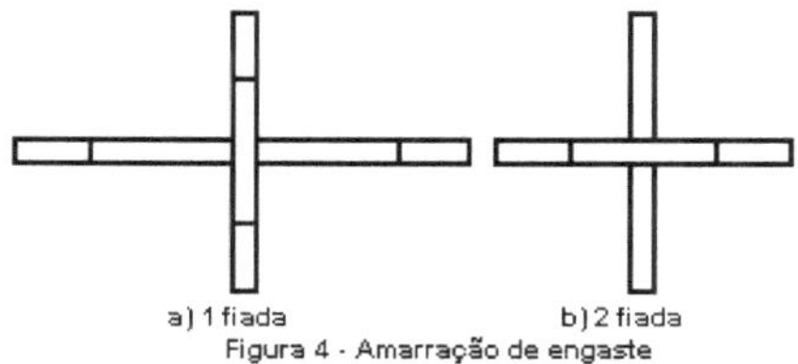


Figure 4 - Crimp fastening

**13.1.7** When joining masonry components to framing elements, additional components must be used to ensure fixing at discrete points, as well as using materials that continuously fill the gaps in the connection.

13.1.8 The gypsum glue must flow through the joints as the blocks are laid and tapped with the rubber hammer. The joints should be a maximum of 2mm thick and continuous.

## 14 Specific conditions

### 14.1 Wet areas

14.1.1 In wet areas, where there is water on the floor, it is recommended to use hydrophilic components in the first row.

14.1.2 In areas subject to capillary water rise, the use of waterproofing devices at the base is recommended. Figure 5 shows some of the possible types of insulation.

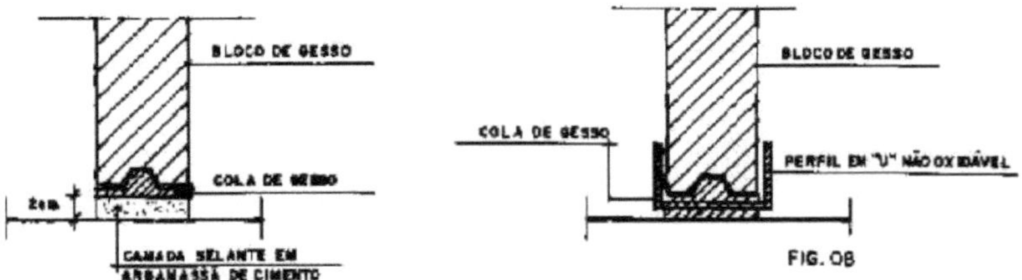


Figure 5 - Possible types of insulation

### 14.2 Weak structures

14.2.1 In masonry built on structural elements that are not very sturdy, it is recommended that devices be used to absorb deformations within the limits set, keeping the masonry stable and free of cracks. Figures 6, 7 and 8 show some of the devices that can be used at the junctions with vertical elements (walls and/or columns) and horizontal elements (floor, slab and beam).

In masonry built on structural elements that are not very sturdy, it is recommended that devices be used to absorb deformations, within the expected limits, keeping the masonry stable and free of cracks. Figures 9, 10 and 11 show some of the devices that can be used at the junctions with vertical elements (walls and/or columns) and horizontal elements (floor, slab and beam).

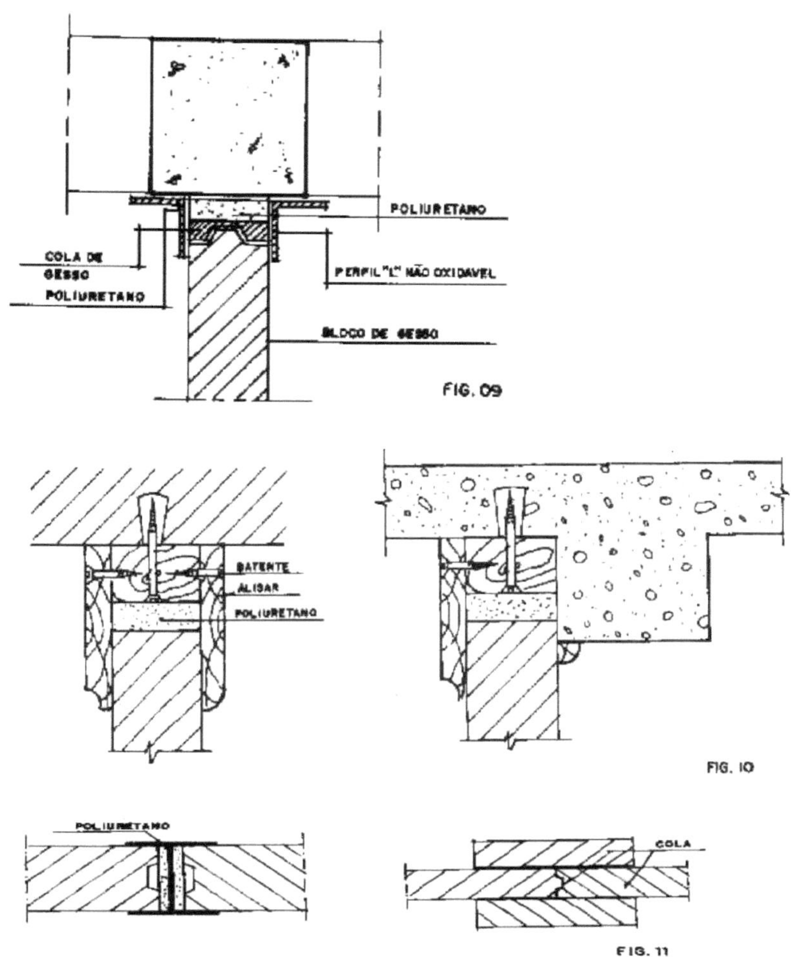


14.2.2 When building masonry in long spans (L > 5m), it is recommended to use structuring devices that guarantee performance in accordance with the specifications of Basic Text 02.002.40.12.

14.2.3 When building masonry > 3m in height, it is recommended to use structuring devices that guarantee performance in accordance with the specifications of Basic Text 02.002.40.12.

### 14.3 Fixing suspended parts

The fixing of suspended parts must comply with NBR 15575

### 14.4 Vergas

14.4.1 Openings, doors and windows > 0.90m must be linteled.

14.4.2 The lintels must exceed the width of the span by at least 20 centimetres on each side and must have a maximum height of 10 centimetres.

14.4.3 When the span is greater than 2.40m, the lintel must be calculated as a beam.

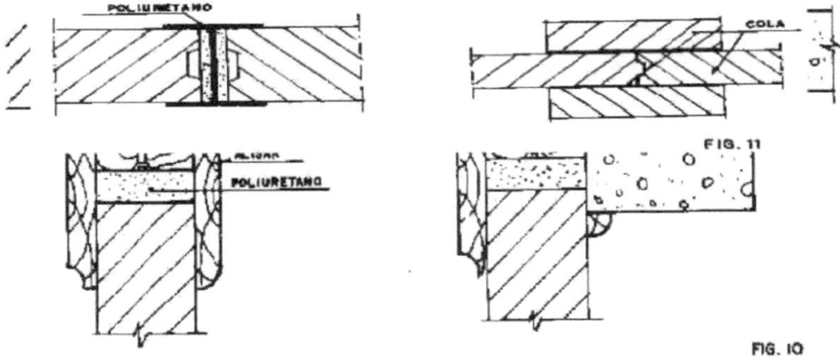


### 14.5 Scaffolding

Scaffolding must meet the requirements of NBR 6494.

### 14.6 Facilities

14.6.1 If it is necessary to make openings in the masonry to embed the installations, these should only be started after the locking has been carried out.

14.6.2 The necessary grooves can be made with a cutting blade or saw, making sure that they do not coincide with the vertical joints.

## 15 Inspection

1.1.1 **1** It is the responsibility of the site supervisor to inspect and take delivery of the masonry.

1.1.2 **2** All masonry must be inspected according to the criteria indicated in this Standard

### 15.2 Rental

15.2.1 It must be checked before the masonry is assembled and verified after the masonry has been assembled, and must be in accordance with the dimensions of the specific project.

15.2.2 Instruments with the precision of tape measures and building squares can be used for this check.

### 15.3 Wall flatness

15.3.1 It must be checked during the assembly of the masonry and verified after the masonry has been assembled, with no distortion greater than 5 mm.

15.3.2 A metal ruler can be used to check this by positioning it at various points on the wall.

### 15.4 Plumb

It must be checked during the assembly of the masonry and verified after the masonry has been assembled.

### 15.5 Level

It must be checked during the assembly of the masonry and verified after the masonry has been assembled. This can be done using a transparent plastic hose with a diameter greater than or equal to 13 mm.

Printed by Books on Demand GmbH, Norderstedt / Germany